NP 265
Edition 2 – 2005

GW00746564

Tidal Stream Atlas

FRANCE
WEST COAST

The Charts

This atlas contains two sets of 13 charts showing tidal streams at hourly intervals commencing 6 hours before HW Brest and ending 6 hours after HW Brest. The times of HW Brest and other details of the prediction for this port are given in NP 202, Admiralty Tide Tables Vol 2, which is published annually. NP 202 also gives tidal predictions for a number of ports and tidal stations in the area covered by this atlas.

On the tidal stream charts the directions are shown by arrows which are graded in weight and, where possible, in length to indicate the approximate strength of the tidal stream. Thus $\longrightarrow$ indicates a weak stream and $\longrightarrow$ indicates a strong stream.

The figures against the arrows give the mean neap and spring rates in tenths of a knot, thus: 19,34 indicates a mean neap rate of 1·9 knots and a mean spring rate of 3·4 knots. The comma indicates the approximate position at which the observations were obtained.

Computation of Rates – Example

Required to predict the rate of the tidal stream in position 48°02′N 4°46′W at 1100 on a day for which the tidal prediction for Brest (extracted from NP 202) is:

0058	1·0m
0700	7·3m
1323	0.8m
1924	6.8m

Mean Range of tide at Brest for the day is 6·1m.

The appropriate chart in the atlas is that for *4 hours after HW Brest* and this gives mean neap and spring rates for the required position of 48°02′N 4°46′W of 15,26 (1·5 kn, 2·6 kn).

Enter the diagram *Computation of Rates* opposite with these mean neap and spring rates, joining the dots representing them with a ruler. From the intersection of this line with the horizontal line representing the range at Brest (6·1m) follow the line vertically to the scale of Tidal Stream Rates (top or bottom) and read off the predicted rate — in this example 2·7 knots.

**Published by the
Hydrographic Office, Taunton,
under the direction of
Dr D W Williams
United Kingdom
National Hydrographer.**

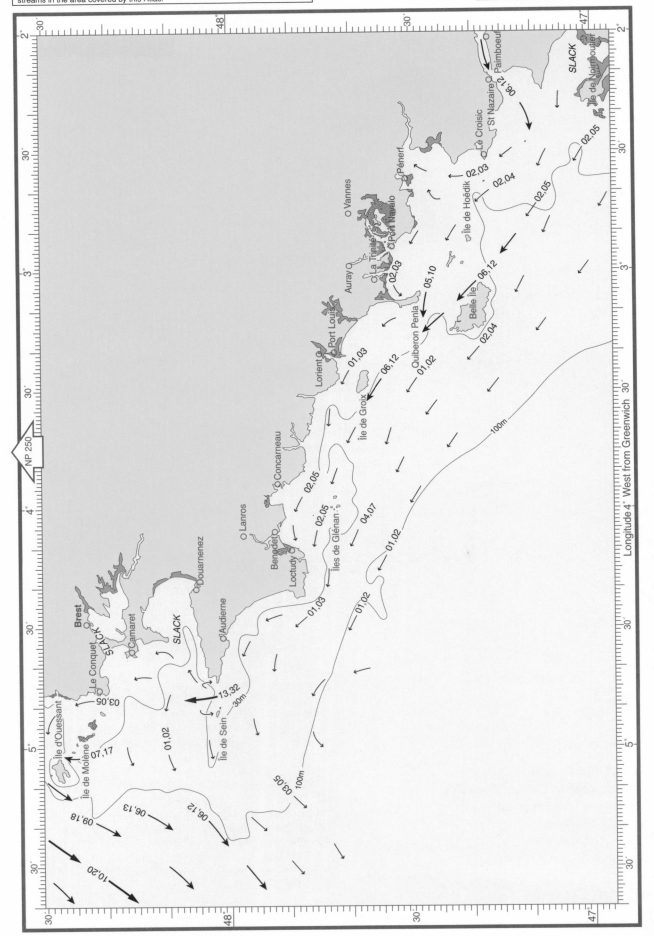

CAUTION:- Due to the very strong rates of tidal streams in some of the areas covered by this Atlas, many eddies and overfalls may occur. Where possible some indication of these has been included. In many areas there is either insufficient information or the eddies are unstable.

Strong winds may also have a great effect on the rate and direction of the tidal streams in the area covered by this Atlas.

NP 250

CAUTION:- Due to the very strong rates of tidal streams in some of the areas covered by this Atlas, many eddies and overfalls may occur. Where possible some indication of these has been included. In many areas there is either insufficient information or the eddies are unstable.
Strong winds may also have a great effect on the rate and direction of the tidal streams in the area covered by this Atlas.

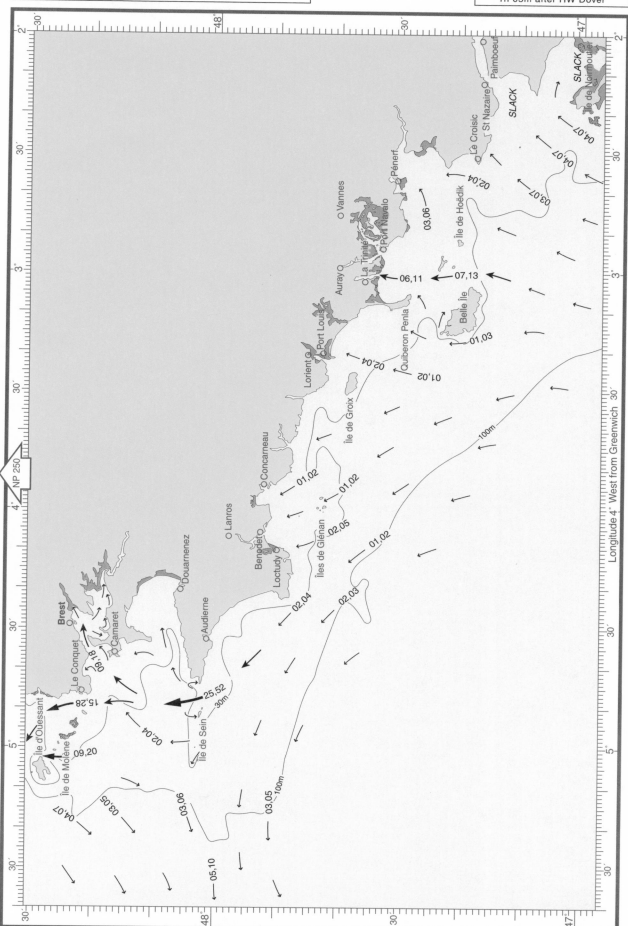

NP 250

CAUTION:- Due to the very strong rates of tidal streams in some of the areas covered by this Atlas, many eddies and overfalls may occur. Where possible some indication of these has been included. In many areas there is either insufficient information or the eddies are unstable.

Strong winds may also have a great effect on the rate and direction of the tidal streams in the area covered by this Atlas.

4

BEFORE HIGH WATER BREST

2h 05m after HW Dover

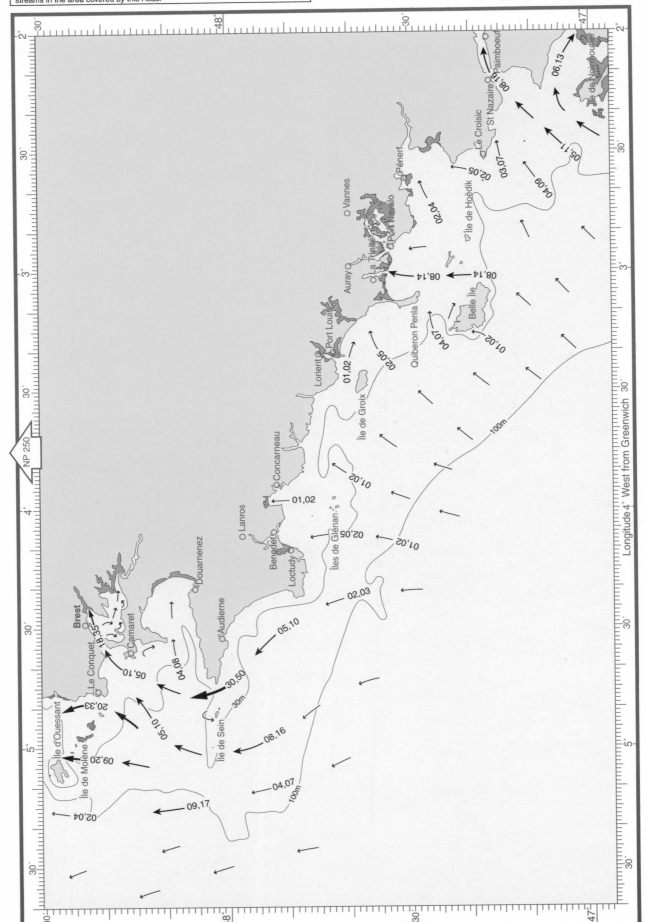

CAUTION:- Due to the very strong rates of tidal streams in some of the areas covered by this Atlas, many eddies and overfalls may occur. Where possible some indication of these has been included. In many areas there is either insufficient information or the eddies are unstable.

Strong winds may also have a great effect on the rate and direction of the tidal streams in the area covered by this Atlas.

3 | **BEFORE HIGH WATER BREST**

3h 05m ater HW Dover

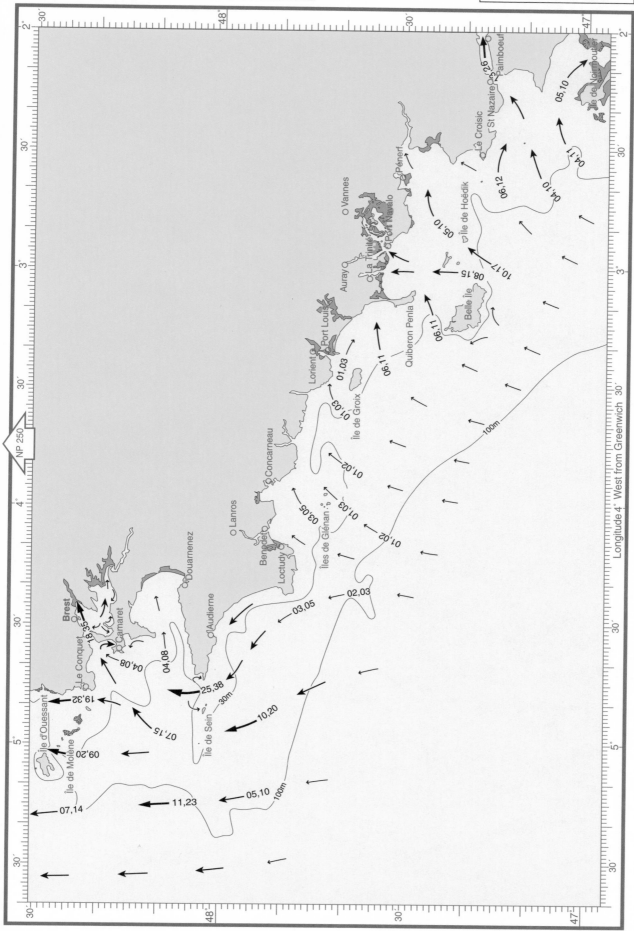

CAUTION:- Due to the very strong rates of tidal streams in some of the areas
covered by this Atlas, many eddies and overfalls may occur. Where possible some
indication of these has been included. In many areas there is either insufficient
information or the eddies are unstable.
Strong winds may also have a great effect on the rate and direction of the tidal
streams in the area covered by this Atlas.

2 | **BEFORE**
HIGH WATER
BREST

4h 05m after HW Dover

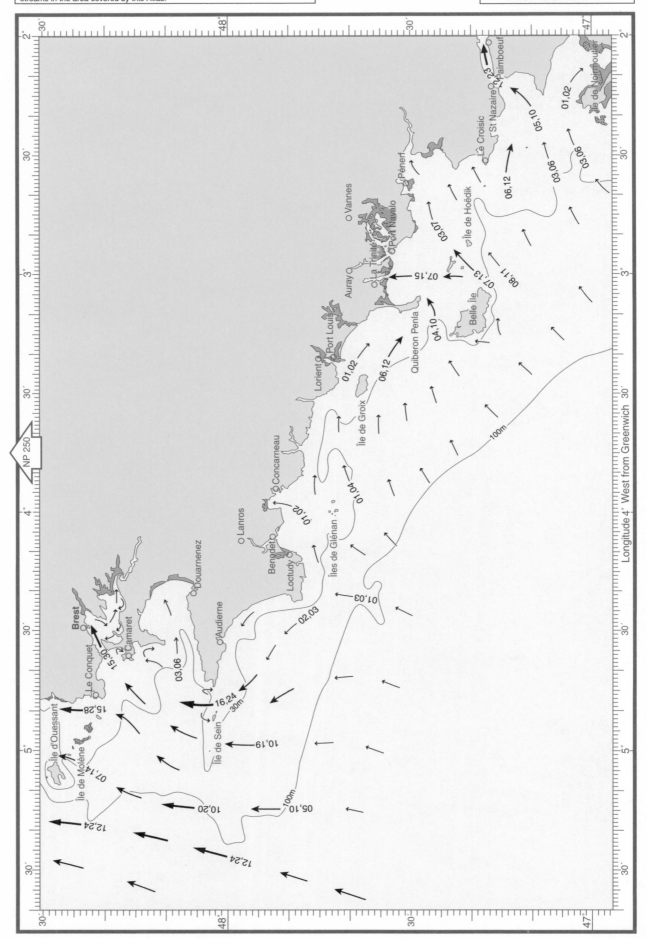

CAUTION:- Due to the very strong rates of tidal streams in some of the areas covered by this Atlas, many eddies and overfalls may occur. Where possible some indication of these has been included. In many areas there is either insufficient information or the eddies are unstable.
Strong winds may also have a great effect on the rate and direction of the tidal streams in the area covered by this Atlas.

NP 250

Longitude 4° West from Greenwich

Brest
Le Conquet
Camaret
île d'Ouessant
île de Molène
île de Sein
Douarnenez
Audierne
Lanros
Concarneau
Bengdet
Loctudy
îles de Glénan
île de Groix
Lorient
Port Louis
Quiberon Penla
Auray
La Trinité
Vannes
Port Navalo
Penerf
Belle île
île de Hoëdik
Le Croisic
St Nazaire
Paimboeuf
île de Noirmoutier

SLACK

09,18
02,05
02,04
03,06
08,16
06,19
04,07
01,04
01,03
06,11
01,02
04,06
03,04
02,05
01,02
01,02
01,0
01,02
03,0
06,07
04,09
04,08
06,11
08,17
13,25
11,22
08,16
09,18

100m

HIGH WATER
BREST

6h 05m after HW DOVER

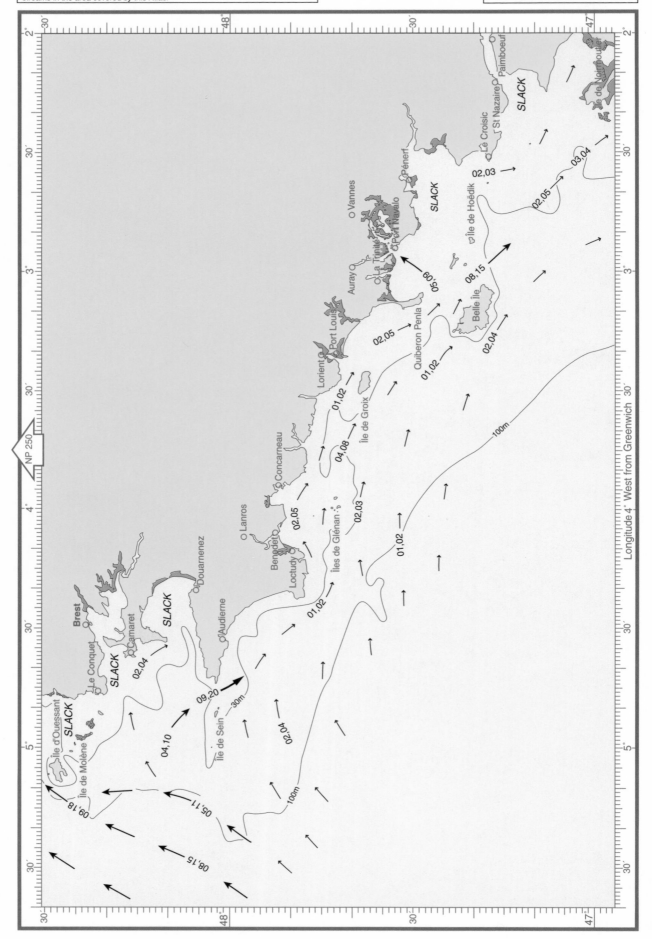

NP 250

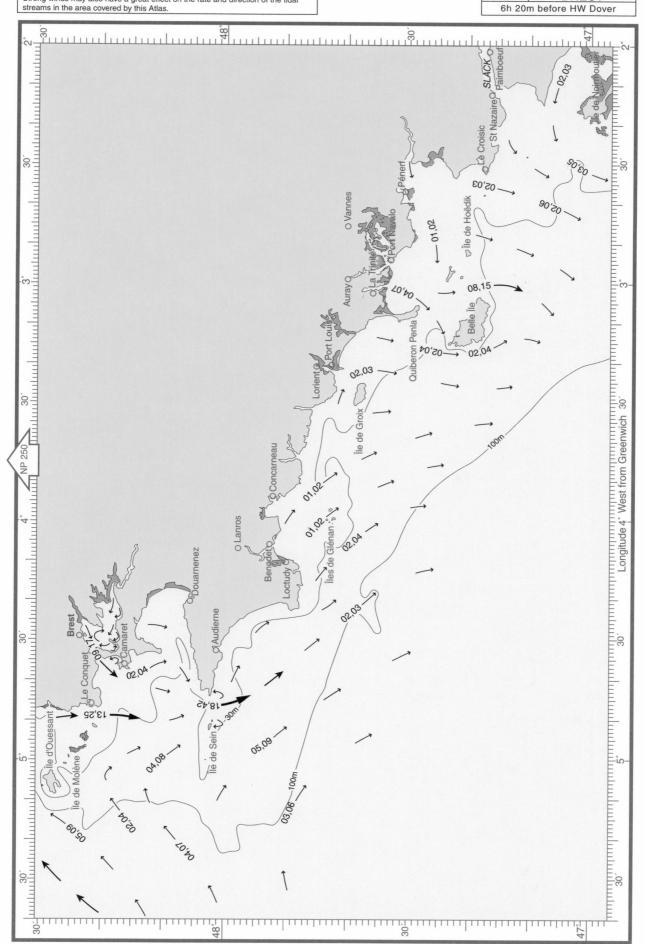

CAUTION:- Due to the very strong rates of tidal streams in some of the areas covered by this Atlas, many eddies and overfalls may occur. Where possible some indication of these has been included. In many areas there is either insufficient information or the eddies are unstable.

Strong winds may also have a great effect on the rate and direction of the tidal streams in the area covered by this Atlas.

2 | **AFTER HIGH WATER BREST**
5h 20m before HW Dover

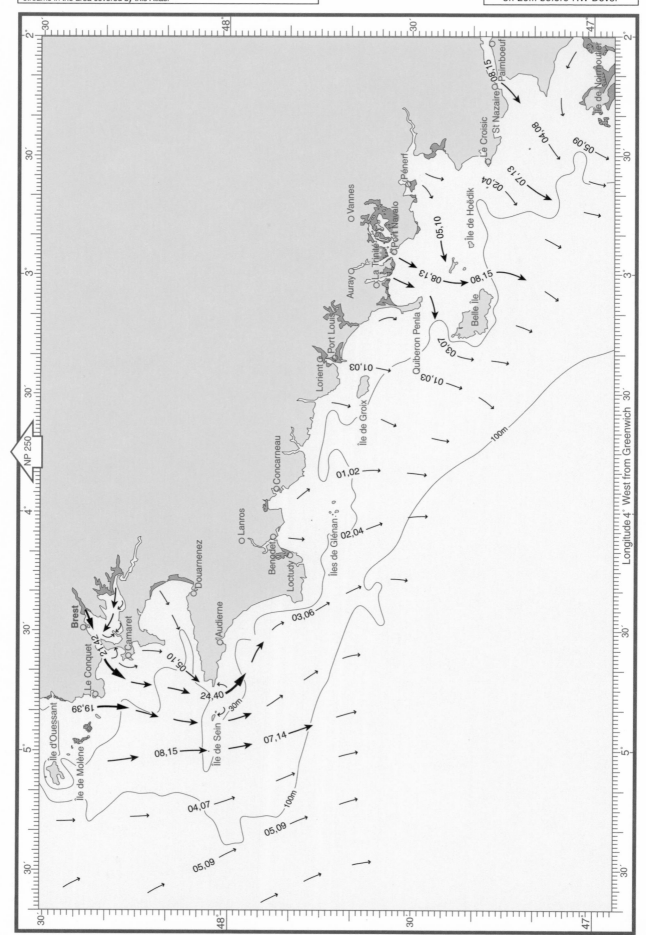

CAUTION:- Due to the very strong rates of tidal streams in some of the areas covered by this Atlas, many eddies and overfalls may occur. Where possible some indication of these has been included. In many areas there is either insufficient information or the eddies are unstable.
Strong winds may also have a great effect on the rate and direction of the tidal streams in the area covered by this Atlas.

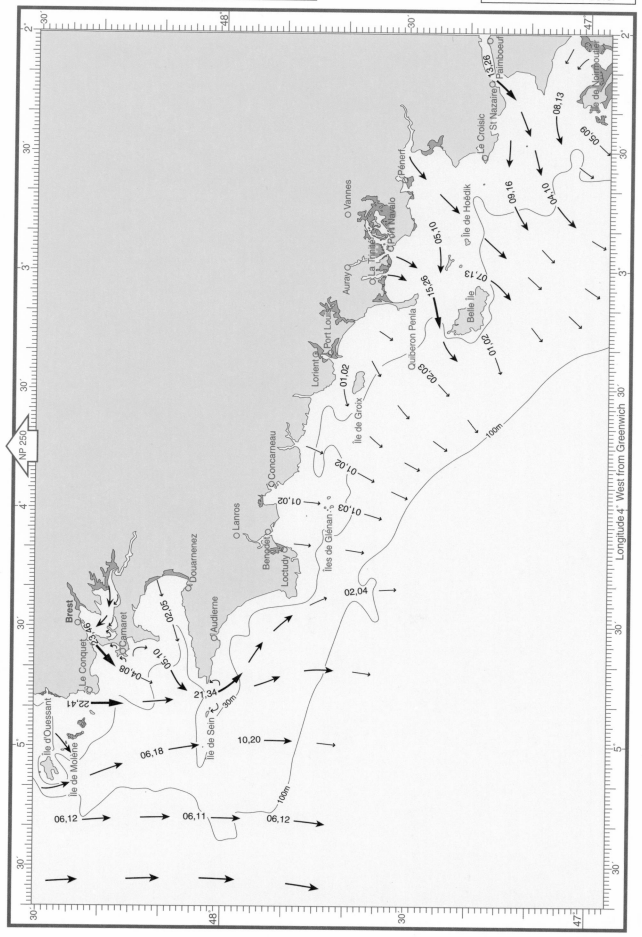

CAUTION:- Due to the very strong rates of tidal streams in some of the areas covered by this Atlas, many eddies and overfalls may occur. Where possible some indication of these has been included. In many areas there is either insufficient information or the eddies are unstable.
Strong winds may also have a great effect on the rate and direction of the tidal streams in the area covered by this Atlas.

4 **AFTER HIGH WATER BREST**
3h 20m before HW Dover

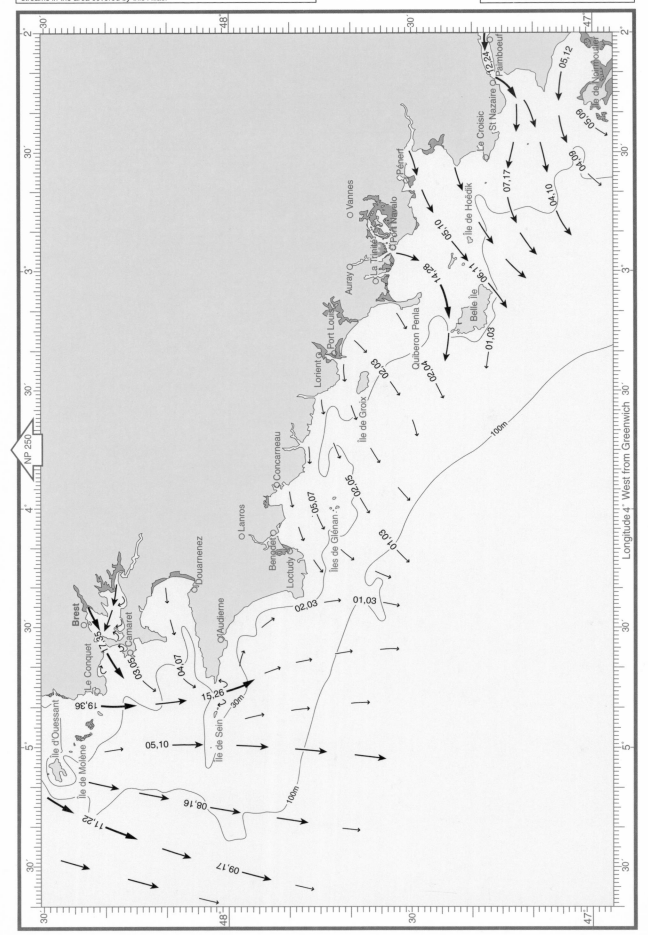

CAUTION:- Due to the very strong rates of tidal streams in some of the areas covered by this Atlas, many eddies and overfalls may occur. Where possible some indication of these has been included. In many areas there is either insufficient information or the eddies are unstable.

Strong winds may also have a great effect on the rate and direction of the tidal streams in the area covered by this Atlas.

5 **AFTER HIGH WATER BREST**

2h 20m before HW Dover

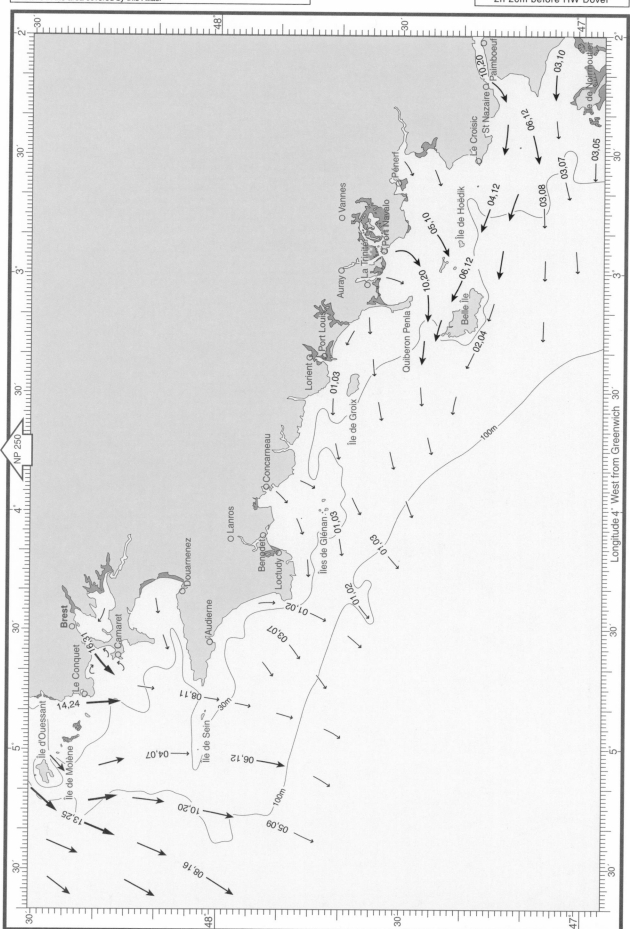

NP 250

CAUTION:- Due to the very strong rates of tidal streams in some of the areas covered by this Atlas, many eddies and overfalls may occur. Where possible some indication of these has been included. In many areas there is either insufficient information or the eddies are unstable.
Strong winds may also have a great effect on the rate and direction of the tidal streams in the area covered by this Atlas.

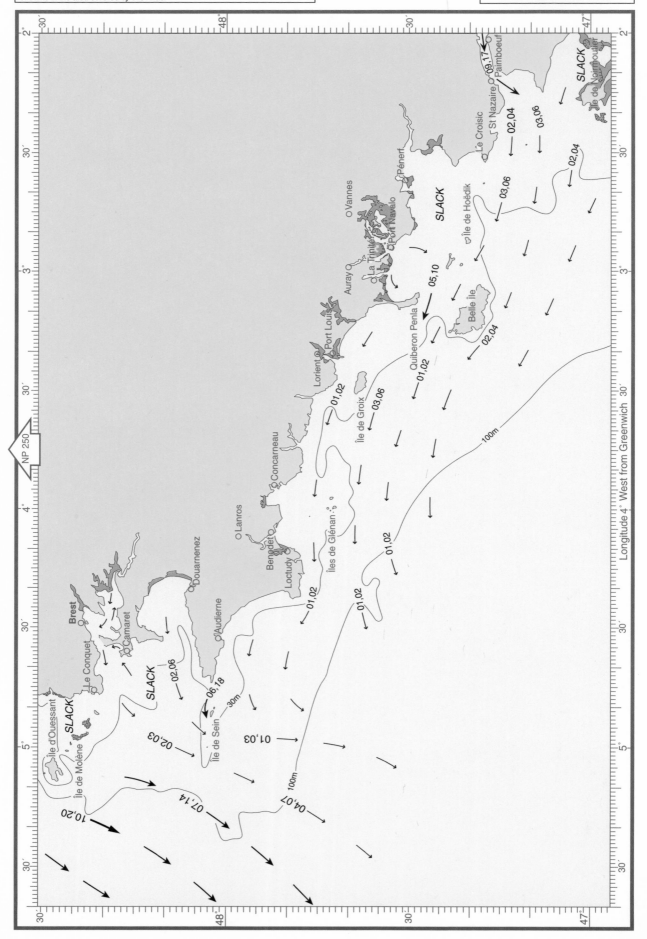

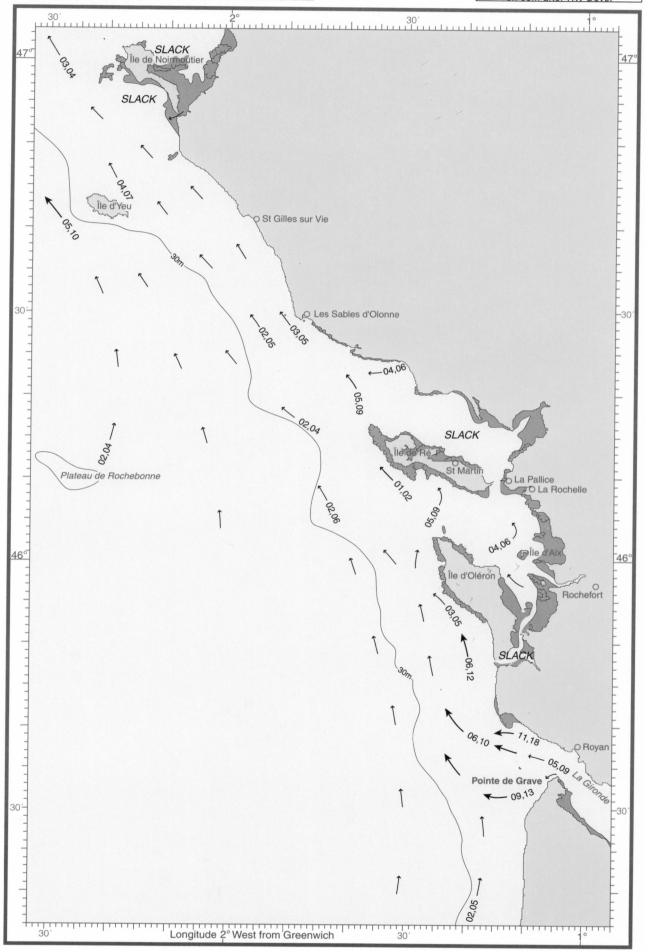

SLACK
Île de Noirmoutier
SLACK
03.04
04.07
Île d'Yeu
05.10
30m
St Gilles sur Vie
02.05
03.05
Les Sables d'Olonne
04,06
05.09
02.04
02.04
Plateau de Rochebonne
SLACK
Île de Ré
St Martin
01,02
La Pallice
La Rochelle
05.09
02.06
04,06
Île d'Aix
Île d'Oléron
Rochefort
03.05
06.12
SLACK
30m
11,18
06,10
Royan
05.09 La Gironde
Pointe de Grave
09,13
02.05

Longitude 2° West from Greenwich

CAUTION:- Due to the very strong rates of tidal streams in some of the areas covered by this Atlas, many eddies and overfalls may occur. Where possible some indication of these has been included. In many areas there is either insufficient information or the eddies are unstable.
Strong winds may also have a great effect on the rate and direction of the tidal streams in the area covered by this Atlas.

5 **BEFORE HIGH WATER BREST**
5h 10m before HW Pointe de Grave
1h 05m after HW Dover

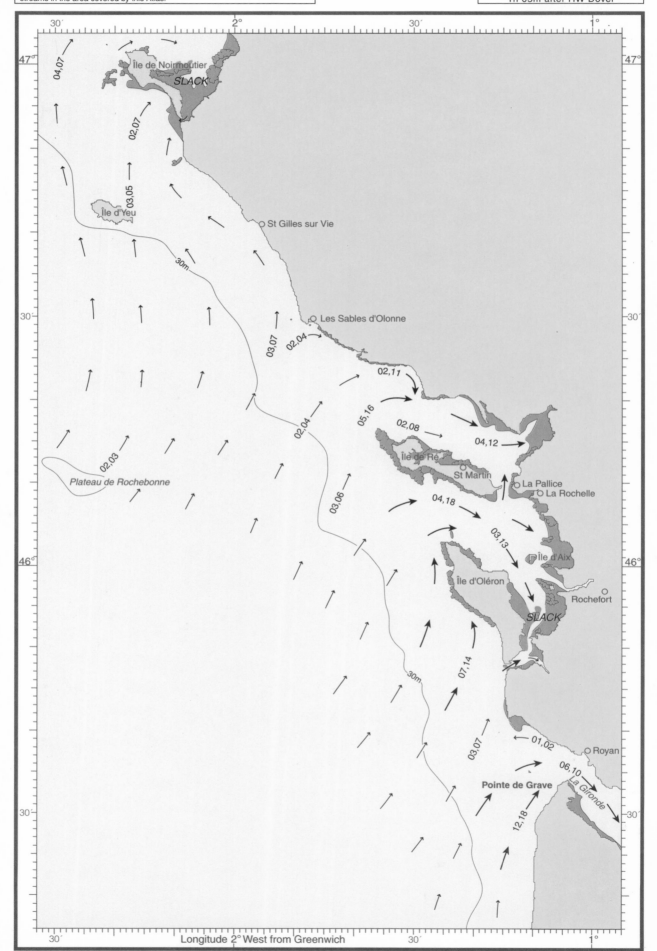

Longitude 2° West from Greenwich

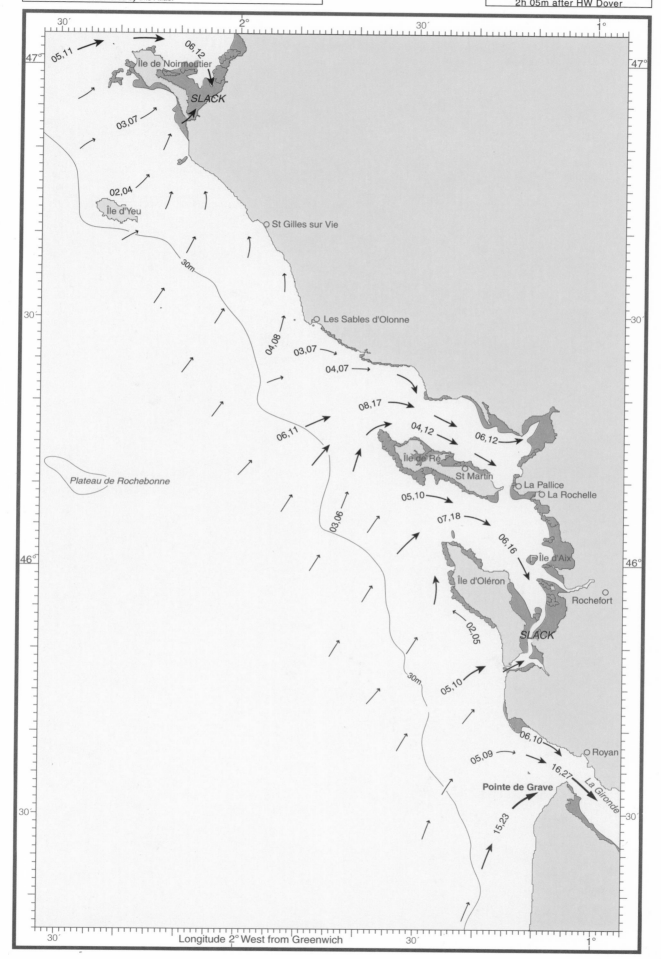

CAUTION:- Due to the very strong rates of tidal streams in some of the areas covered by this Atlas, many eddies and overfalls may occur. Where possible some indication of these has been included. In many areas there is either insufficient information or the eddies are unstable.
Strong winds may also have a great effect on the rate and direction of the tidal streams in the area covered by this Atlas.

4 **BEFORE HIGH WATER BREST**
4h 10m before HW Pointe de Grave
2h 05m after HW Dover

47°

30′

05,11

06,12

Île de Noirmoutier

SLACK

03,07

02,04

Île d'Yeu

30m

St Gilles sur Vie

30′

Les Sables d'Olonne

04,08

03,07

04,07

08,17

06,11

04,12

06,12

Île de Ré

St Martin

La Pallice

La Rochelle

Plateau de Rochebonne

05,10

07,18

06,16

Île d'Aix

03,06

46°

Île d'Oléron

Rochefort

02,05

SLACK

30m

05,10

06,10

Royan

05,09

16,27

La Gironde

Pointe de Grave

15,23

30′

Longitude 2° West from Greenwich   30′   1°

CAUTION:- Due to the very strong rates of tidal streams in some of the areas covered by this Atlas, many eddies and overfalls may occur. Where possible some indication of these has been included. In many areas there is either insufficient information or the eddies are unstable.
Strong winds may also have a great effect on the rate and direction of the tidal streams in the area covered by this Atlas.

3 **BEFORE**
HIGH WATER
BREST
3h 10m before HW Pointe de Grave
3h 05m after HW Dover

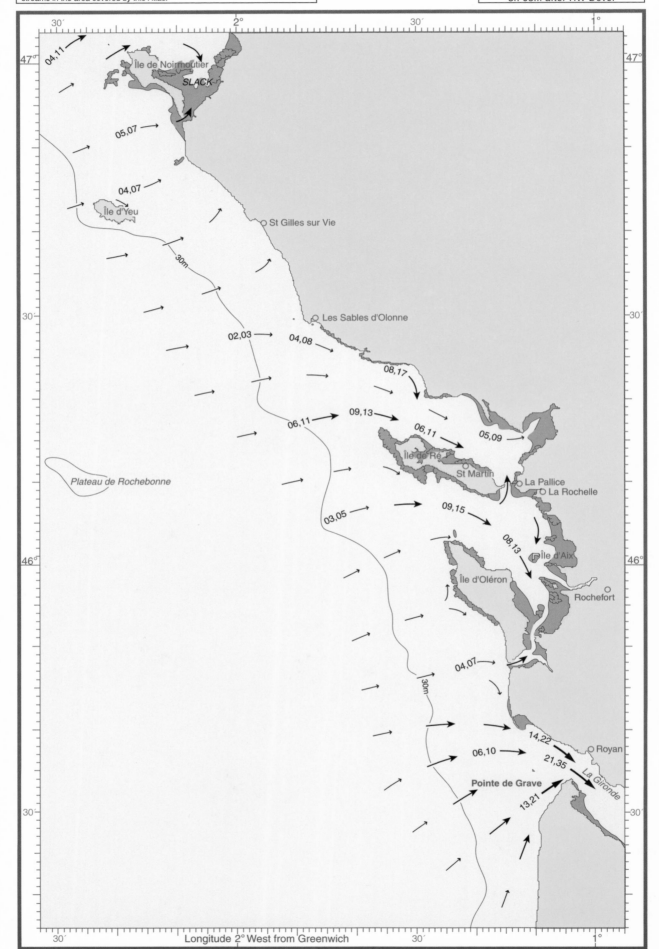

CAUTION:- Due to the very strong rates of tidal streams in some of the areas covered by this Atlas, many eddies and overfalls may occur. Where possible some indication of these has been included. In many areas there is either insufficient information or the eddies are unstable.
Strong winds may also have a great effect on the rate and direction of the tidal streams in the area covered by this Atlas.

2 **BEFORE HIGH WATER BREST**
2h 10m before HW Pointe de Grave
4h 05m after HW Dover

03,07

Île de Noirmoutier

SLACK

04,04

03,07

Île d'Yeu

St Gilles sur Vie

30m

Les Sables d'Olonne

05,08

04,07

08,12

05,09

09,13

06,11

04,06

Plateau de Rochebonne

Île de Ré

St Martin

La Pallice

La Rochelle

02,04

07,11

07,07

Île d'Aix

Île d'Oléron

Rochefort

03,06

30m

16,26

05,11

Royan

19,31

La Gironde

Pointe de Grave

10,16

Longitude 2° West from Greenwich

CAUTION:- Due to the very strong rates of tidal streams in some of the areas covered by this Atlas, many eddies and overfalls may occur. Where possible some indication of these has been included. In many areas there is either insufficient information or the eddies are unstable.
Strong winds may also have a great effect on the rate and direction of the tidal streams in the area covered by this Atlas.

1	BEFORE HIGH WATER BREST

1h 10m before HW Pointe de Grave
5h 05m after HW Dover

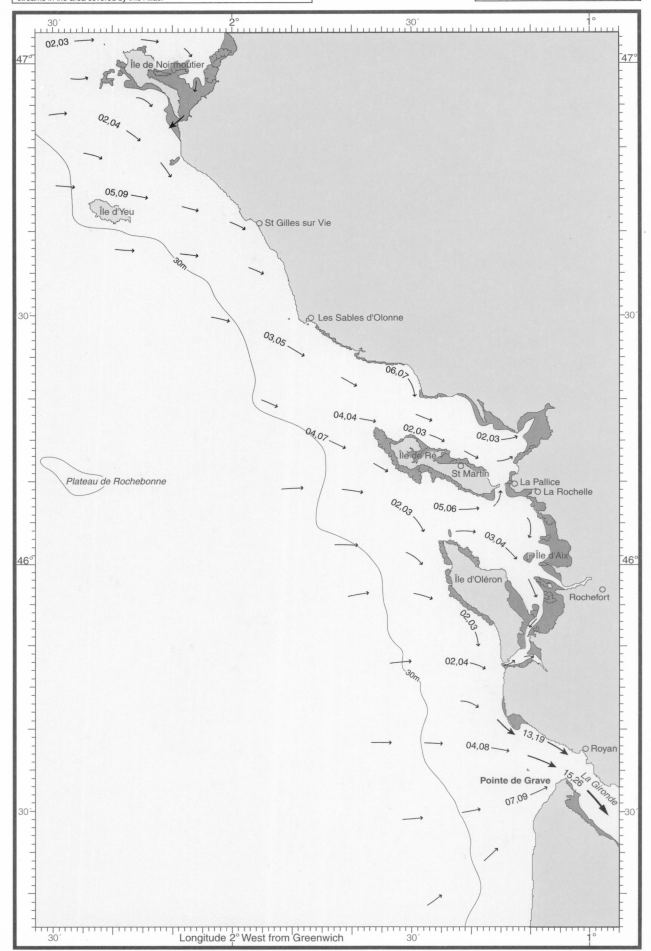

HIGH WATER
BREST
10m before HW Pointe de Grave
6h 05m after HW Dover

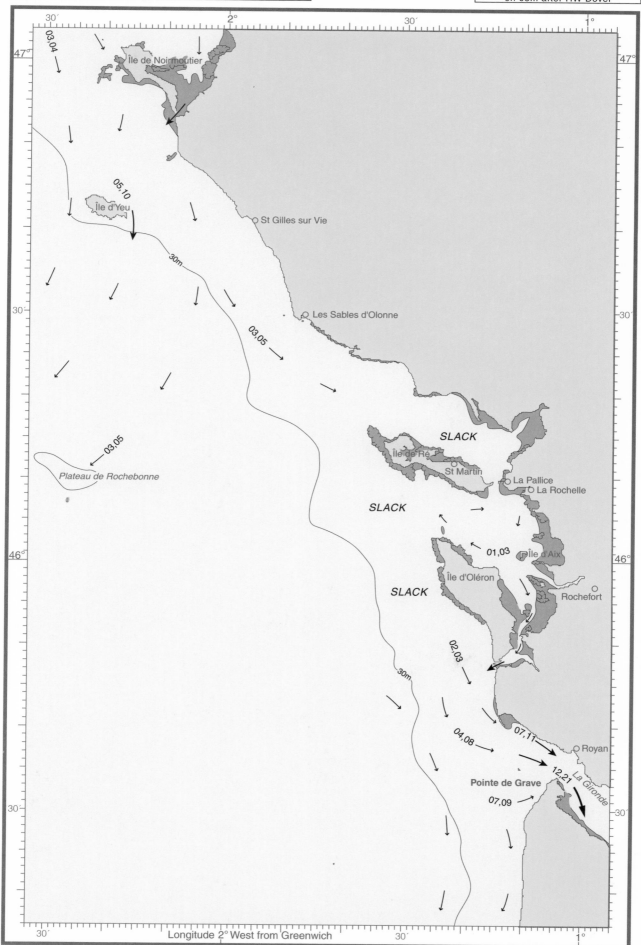

Île de Noirmoutier

03,04

05,10
Île d'Yeu

30m

St Gilles sur Vie

Les Sables d'Olonne

03,05

03,05
Plateau de Rochebonne

SLACK
Île de Ré
St Martin
La Pallice
La Rochelle

SLACK

SLACK
01,03
Île d'Aix
Île d'Oléron
Rochefort

30m

02,03

04,08
07,11
Royan
Pointe de Grave
12,21
La Gironde
07,09

Longitude 2° West from Greenwich

CAUTION:- Due to the very strong rates of tidal streams in some of the areas covered by this Atlas, many eddies and overfalls may occur. Where possible some indication of these has been included. In many areas there is either insufficient information or the eddies are unstable.
Strong winds may also have a great effect on the rate and direction of the tidal streams in the area covered by this Atlas.

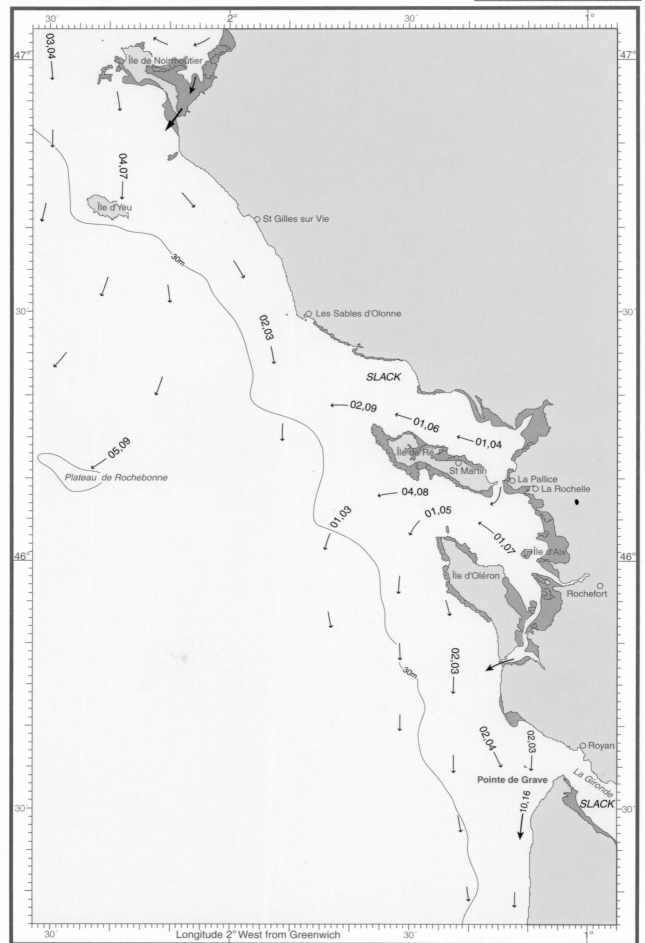

Île de Noirmoutier

03,04

04,07

Île d'Yeu

30m

St Gilles sur Vie

02,03

Les Sables d'Olonne

SLACK

02,09

01,06

01,04

Île de Ré

St Martin

La Pallice

La Rochelle

05,09

Plateau de Rochebonne

04,08

01,05

01,07

Île d'Aix

01,03

Île d'Oléron

Rochefort

30m

02,03

02,04

02,03

Royan

La Gironde

10,16

Pointe de Grave

SLACK

Longitude 2° West from Greenwich

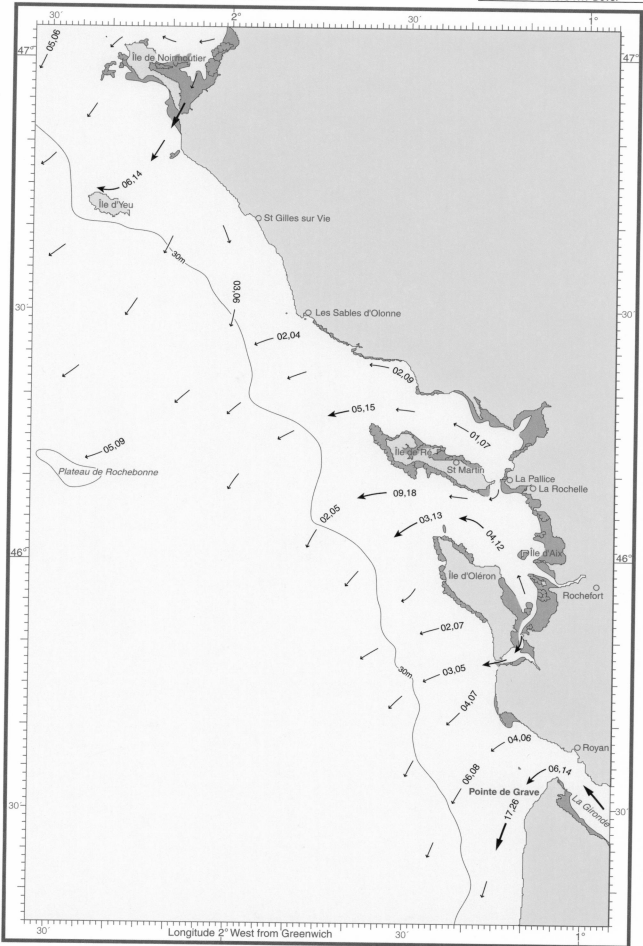

CAUTION:- Due to the very strong rates of tidal streams in some of the areas covered by this Atlas, many eddies and overfalls may occur. Where possible some indication of these has been included. In many areas there is either insufficient information or the eddies are unstable.
Strong winds may also have a great effect on the rate and direction of the tidal streams in the area covered by this Atlas.

Longitude 2° West from Greenwich

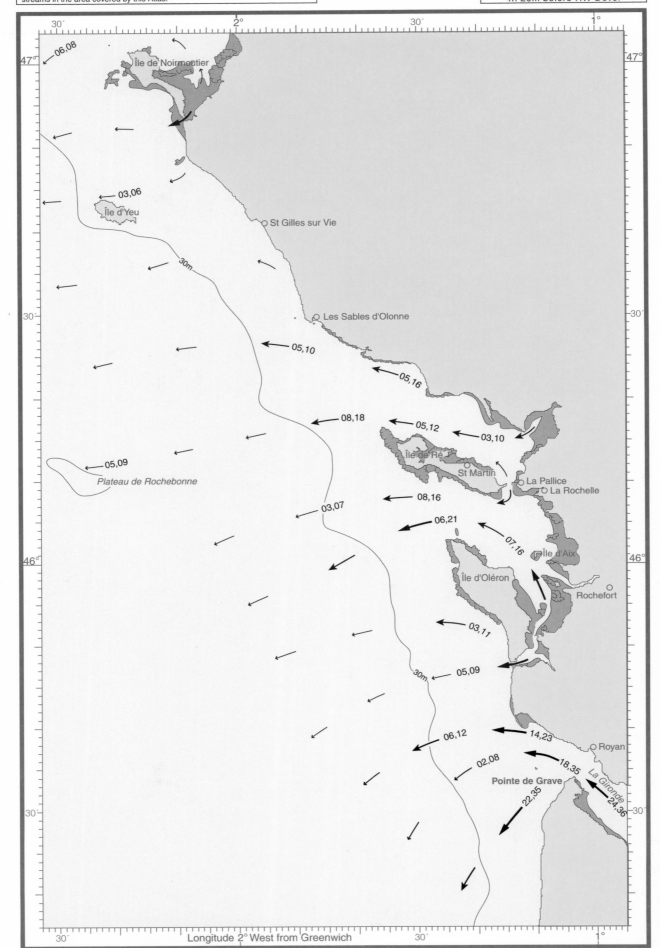

CAUTION:- Due to the very strong rates of tidal streams in some of the areas covered by this Atlas, many eddies and overfalls may occur. Where possible some indication of these has been included. In many areas there is either insufficient information or the eddies are unstable.
Strong winds may also have a great effect on the rate and direction of the tidal streams in the area covered by this Atlas.

3 | **AFTER HIGH WATER BREST**
2h 50m after HW Pointe de Grave
4h 20m before HW Dover

Île de Noirmoutier

06,08

03,06

Île d'Yeu

St Gilles sur Vie

30m

Les Sables d'Olonne

05,10

05,16

08,18

05,12

03,10

Île de Ré

05,09

Plateau de Rochebonne

St Martin

La Pallice

La Rochelle

08,16

06,21

07,16

Île d'Aix

03,07

46°

Île d'Oléron

Rochefort

03,11

30m

05,09

06,12

14,23

Royan

02,08

18,35

La Gironde

Pointe de Grave

22,35

24,36

Longitude 2° West from Greenwich

CAUTION:- Due to the very strong rates of tidal streams in some of the areas covered by this Atlas, many eddies and overfalls may occur. Where possible some indication of these has been included. In many areas there is either insufficient information or the eddies are unstable.
Strong winds may also have a great effect on the rate and direction of the tidal streams in the area covered by this Atlas.

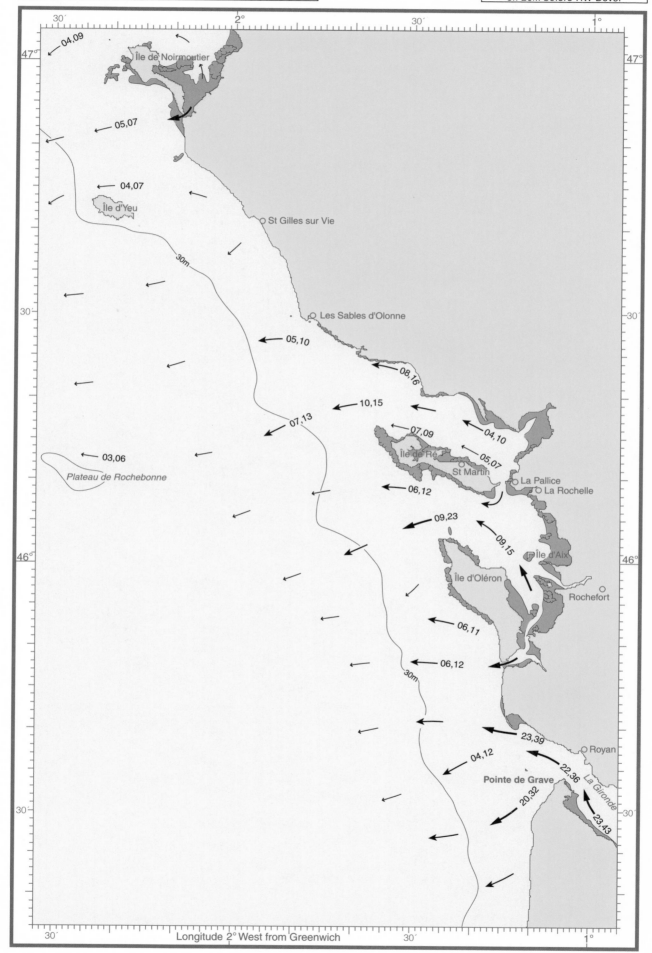

Longitude 2° West from Greenwich

CAUTION:- Due to the very strong rates of tidal streams in some of the areas covered by this Atlas, many eddies and overfalls may occur. Where possible some indication of these has been included. In many areas there is either insufficient information or the eddies are unstable.

Strong winds may also have a great effect on the rate and direction of the tidal streams in the area covered by this Atlas.

5 | **AFTER HIGH WATER BREST**

4h 50m after HW Pointe de Grave
2h 20m before HW Dover

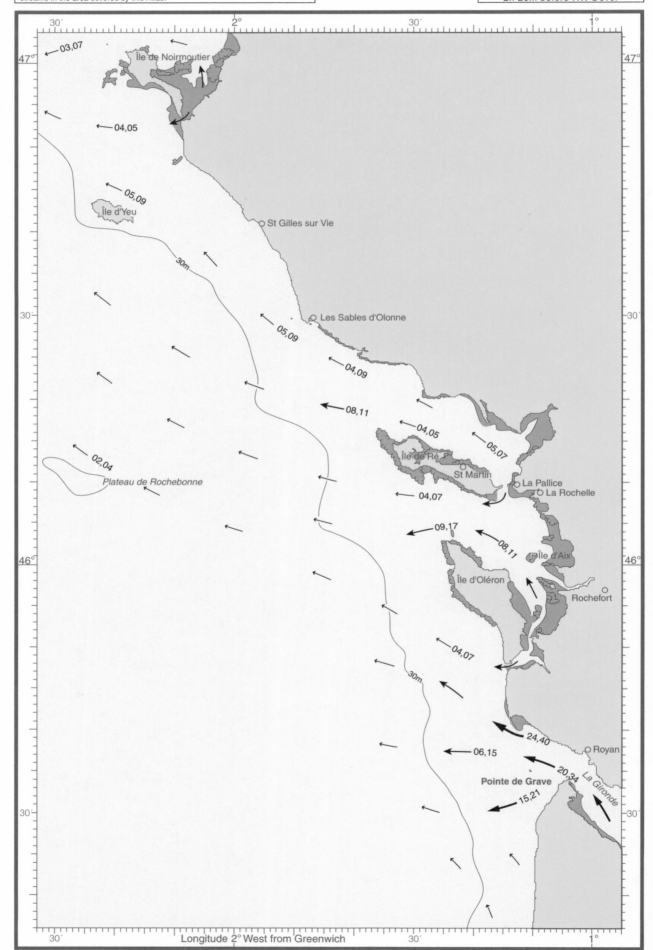

CAUTION:- Due to the very strong rates of tidal streams in some of the areas covered by this Atlas, many eddies and overfalls may occur. Where possible some indication of these has been included. In many areas there is either insufficient information or the eddies are unstable.
Strong winds may also have a great effect on the rate and direction of the tidal streams in the area covered by this Atlas.

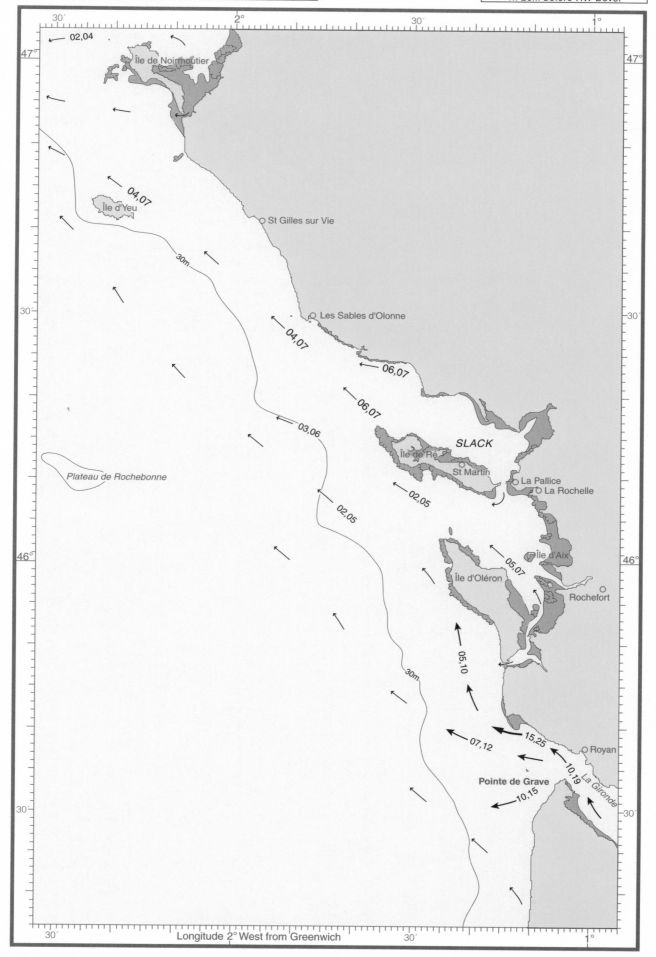

Île de Noirmoutier
02,04
04,07
Île d'Yeu
St Gilles sur Vie
30m
Les Sables d'Olonne
04,07
06,07
06,07
SLACK
03,06
Île de Ré
St Martin
La Pallice
La Rochelle
Plateau de Rochebonne
02,05
02,05
Île d'Aix
05,07
Île d'Oléron
Rochefort
05,10
30m
07,12
15,25
Royan
Pointe de Grave
La Gironde
10,19
10,15

Longitude 2° West from Greenwich

ADMIRALTY
TIDAL PUBLICATIONS

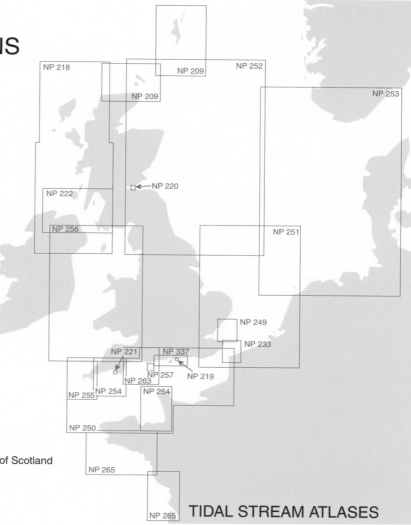

TIDAL STREAM ATLAS - NW Europe

NP 209 Orkney and Shetland Islands
NP 218 North Coast of Ireland and West Coast of Scotland
NP 219 Portsmouth Harbour and Approaches
NP 220 Rosyth Harbour and Approaches
NP 221 Plymouth Harbour and Approaches
NP 222 Firth of Clyde and Approaches
NP 233 Dover Strait
NP 249 Thames Estuary (with co-tidal charts)
NP 250 The English Channel
NP 251 North Sea, Southern Part
NP 252 North Sea, North Western Part
NP 253 North Sea, Eastern Part
NP 254 The West Country, Falmouth to Teignmouth
NP 255 Falmouth to Padstow, including the Isles of Scilly
NP 256 Irish Sea and Bristol Channel
NP 257 Approaches to Portland
NP 263 Lyme Bay
NP 264 The Channel Islands and adjacent Coast of France
NP 265 France, West Coast
NP 337 The Solent and adjacent waters

ADMIRALTY TIDE TABLES (published annually)

NP 201 Volume 1 : United Kingdom and Ireland (including European Channel Ports)
NP 202 Volume 2 : Europe (excluding United Kingdom and Ireland), Mediterranean Sea and Atlantic Ocean
NP 203 Volume 3 : Indian Ocean and South China Sea (including Tidal Stream Tables)
NP 204 Volume 4 : Pacific Ocean (including Tidal Stream Tables)

MISCELLANEOUS TIDAL PUBLICATIONS

NP 160   Tidal Harmonic Constants (European Waters)
NP 164   Dover, times of high water and mean ranges (published annually)
Chart 5057 Dungeness to Hoek van Holland, co-tidal and co-range chart
Chart 5058 British Isles and adjacent waters, co-tidal and co-range lines
Chart 5059 Southern North Sea, co-tidal and co-range chart

DIGITAL TIDAL PUBLICATIONS

DP 550 TotalTide CD-ROM
DP 560 Simplified Harmonic Method of Tidal Prediction (SHM for Windows) CD-ROM

The publications listed above may be obtained from Agents for the sale of Admiralty Charts (listed in Admiralty Notice to Mariners No 2 issued annually)
Prices of tidal atlases, charts and publications are given in NPs 131 and 109 Catalogues of Admiralty Charts and other Hydrographic Publications (published annually)

Predictions of High Water Times

	Lat N	Long W	Approximate time of HW relative to HW at Dover
Ouessant - Baie de Lampaul	48 27	5 06	+0610
Île de Molène	48 24	4 58	+0615
Le Conquet	48 22	4 47	+0605
Brest	48 23	4 30	+0605
Camaret	48 16	4 36	+0555
Morgat	48 13	4 30	+0555
Douarnenez	48 06	4 19	+0555
Île de Sein	48 02	4 51	+0600
Audierne	48 01	4 33	+0535
Le Guilvinec	47 48	4 17	+0545
Pont l'Abbe River - Loctudy	47 50	4 10	+0545
Odet River - Benodet	47 53	4 07	+0555
Corniguel	47 58	4 06	+0615
Concarneau	47 52	3 55	+0545
Île de Penfret	47 44	3 57	+0545
Port Louis	47 42	3 21	+0555
Lorient	47 45	3 21	+0555
Port Tudy	47 39	3 27	+0550
Port-Haliguen	47 29	3 06	+0600
Belle-Île - Le Palais	47 21	3 09	+0555
Crac'h River - La Trinité	47 35	3 01	+0605
Morbihan - Port Navalo	47 33	2 55	+0615
Auray	47 40	2 59	+0630
Vannes	47 40	2 46	+0815
Île de Hoëdic	47 20	2 52	+0550
Pénerf	47 31	2 37	+0605
Le Croisic	47 18	2 31	+0555
Le Pouliguen	47 17	2 25	+0600
La Loire - Le Grande-Charpentier	47 13	2 19	+0550
St Nazaire	47 16	2 12	+0600
Pornic	47 06	2 07	+0555
Île de Noirmoutier - L'Herbaudiere	47 02	2 18	+0555
Fromentine	46 54	2 10	+0555
Île d'Yeu - Port Joinville	46 44	2 21	+0550
St Gilles sur Vie	46 41	1 56	+0555
Les Sables d'Olonne	46 30	1 48	+0555
Île de Ré - St Martin	46 12	1 12	+0600
La Pallice	46 10	1 13	+0605
La Rochelle	46 09	1 09	+0605
Île d'Aix	46 01	1 10	+0600
La Charente - Rochefort	45 57	0 58	+0630
Le Chapus	45 51	1 11	+0600
La Cayenne	45 47	1 08	+0625
Pointe de Gatseau	45 48	1 14	+0615
La Gironde - Royen	45 37	1 00	+0610
Pointe de Grave	45 34	1 04	+0615

Note - These differences applied to Dover predictions in GMT will give predictions for the port in Zone Time - 0100 (MET).

For more accurate predictions, see Admiralty Tide Tables, Volume 2.